Contents

The Glen Canyon Dam in Arizona, USA.

1 Why build a dam?

Windsurfing, like many leisure sports, relies on the availability of large bodies of water.

Water is our most precious **resource**. Nothing can live without it. We need it to grow crops for food and for animals. We need water to drink, to make many things, and to do washing and cooking.

We even use water for **leisure activities** such as sailing and swimming.

But there has to be the right amount of water in the right place at the right time. If there is too much, there are floods. If there is too little, there is **drought**. In both cases plants, animals and even people may die.

This is why dams are so important. Dams are built across rivers to control the flow of water. This is done by means of gates or **sluices** which can be opened to let water through, or shut to hold water back. In this way dams help to stop rivers drying up, and can prevent flooding. River water rises behind dams, forming **artificial** lakes called reservoirs. A reservoir is a reserve of water, saved up to supply houses, schools, factories, farm animals and crops. Dams can also use the power of falling water to generate electricity, which is called hydro-electric power.

DAMS IN HISTORY

For centuries, people built and improved the design of dams, by learning from builders before them.

STRUCTURES

DAMS

Andrew Dunn

Wayland

Titles in this series

Bridges
Dams
Skyscrapers
Tunnels

Words that appear in the glossary are printed in **bold** type the first time they appear in the text.

Editor: Kathryn Smith
Series designer: Joyce Chester

First published in 1993 by
Wayland (Publishers) Limited
61 Western Road, Hove
East Sussex BN3 1JD, England

British Library Cataloguing in Publication Data
Dunn, Andrew
  Dams. — (Structures Series)
  I. Title II. Series
  627
ISBN 0-7502-0497-4

Typeset by Key Origination, Eastbourne, Sussex
Printed by G. Canale & C.S.p.A., Turin

Picture Acknowledgements

Ardea 3; Aspect, 8, 20 (top); Cephas 14 (N. Blythe); Bruce Coleman 11, 22 (J. Foott); Ecoscene 29 (Towse); Eye Ubiquitous 6, 17, (Gerald Cubitt), 28 (A. Carroll); Hutchinson 10 (R. Francis), 24 Kobal 25; Photri 16; Tony Stone *cover* 4, 5 (W. Jacobs), 7, 13, (I. Murphy), 18 (C. Condida), 19 (S. Jauncey), 20 and 21 (bottom) Ed Protchard, 21 (top), 23 (S. Stone); Zefa 9 (B.F. Peterson). Artwork by Steve Wheele.

Drought occurs when a period of fifteen days or more passes without any rain. Desert regions, such as the Sahara in Algeria, pictured here, may go for many years without a single drop of rain.

The earliest dams were simple banks of earth or stones, built for crop **irrigation**.

The Ancient Egyptians relied on the River Nile for almost all their water, so they started building dams to store it for their use.

The oldest dam still in use was built in what is now called Syria, nearly 3,300 years ago.

Ancient dams were also built in Sri Lanka, India and China. The Romans were superb **engineers** too.

This man is standing on the present Marib Dam in Yemen. The original dam was built over 3,000 years ago on the same spot. When the first dam burst, it was an international catastrophe, recorded in both the Koran and the Bible.

Concrete dams like this one in Switzerland are constructed so that they can stand the weight of colossal amounts of water.

They built long roads, bridges and **aqueducts** to supply water across their empire. They also built dams, several of which are still in use today.

Centuries later, engineers began to use the laws of science to build bigger, stronger and safer dams. About 400 years ago Spanish engineers discovered that a curved dam, shaped like an arch on its side, is much stronger than a straight wall, so the walls could be much thinner.

Since then, engineers worldwide have built massively strong dams, bigger and higher than before, often in remote, difficult places. Modern dams can hold back colossal amounts of water, in reservoirs hundreds of kilometres long.

USES OF WATER

How many different ways of using water can you think of? Can you find out how many litres of water are needed to fill up a kitchen sink, a bath, or a glass of water? How much water is used each time a lavatory is flushed? How much is used when you take a shower?

Think of things done or made using water. Make a list, and try to find out how much water each one uses.

2 What's in a dam?

All dams are wider at the bottom than at the top. This is because the water at the bottom of a lake or river is squashed down by the water on top, so that it is under more pressure. The deeper the water, the stronger the dam has to be. Four main types of dam are used.

The Ataturk Dam in Turkey is the world's third largest earth, or embankment dam.

Embankment or Gravity Dam

Buttress dams have triangular-shaped wedges which support a straight concrete wall against the flow of the water.

Buttress Dam

These include embankments of earth and rock, and the **concrete** dams: **gravity**, arch, and **buttress**.

Embankment dams are the simplest and oldest type of dam. They are also the biggest. They may be banks of sand or earth, pressed down by heavy machinery. The sloping side facing the oncoming water is lined with stones and rocks to prevent the soil being washed away. An embankment must be broad, strong and heavy to be safe.

Some embankment dams are made with rocks instead of soil. A *rockfill* dam is heavier than an earth dam, so it need not be as broad. But a layer of clay or concrete may be needed, to stop leaks through gaps in the rocks.

Concrete Dam

Concrete dams are much thinner than embankment dams, because concrete is stronger and heavier than earth and rock. A *gravity dam* is rather like an embankment, but usually the side that faces the oncoming water is straight. It supports the water with its own sheer weight.

Buttress dams can just be straight concrete walls, held up by triangular buttresses. The buttresses **brace** the dam firmly against the valley floor.

Concrete arch dams are very strong, and are particularly suitable for steep, narrow valleys. The walls can be as thin as 3 m wide and still hold back deep reservoirs. The strength of the arch dam lies in its shape. It is like an arch bridge lying on its side, pointing into the water. The weight of the water is spread around the curve of the arch until it meets the sides of the valley. There the arch is supported by solid **foundations**, which anchor it firmly. In fact, the pressure of the water actually helps an arch dam. It pushes against the concrete, **compressing** it.

The Hoover Dam, built from concrete between 1931 and 1935, forms Lake Mead, which is the largest artificial reservoir in the USA.

This makes it even stronger. It also squashes the dam against the valley sides so firmly that it cannot move. A *cupola dam* is an arch dam which also curves from top to bottom.

HOW DAMS ARE BUILT

When engineers are planning a dam, they have to think about several questions. What is the dam for? Where should it be built? And what sort of dam should it be?

If the dam is to provide a water supply, for a town or for irrigating crops, then it should be built where it will form a large reservoir.

To build a dam, the river must often be diverted to a temporary route. This keeps the building site dry.

If it is for generating hydroelectric power, then it should be high up where the force of falling water is strongest. If it is to prevent flooding, it should be built above where the flooding happens.

The engineers then look at the site. How good are the foundations? Is the rock or soil waterproof, or will it leak? They use **geological surveys** to find the answers, and models and computers to find out how much water the dam will have to support, and how big and strong it must be. Then they can decide what sort of dam to design.

THE PRESSURE OF WATER

This simple experiment demonstrates how water can exert pushing pressure.

What you need: Two identical syringes and a thin piece of plastic tube.

What you do: Push the plunger in on syringe A. Fill syringe B with water, tapping it gently to get rid of any bubbles. Fill the plastic tube with water, keeping a finger over one end. Insert the end of syringe A into the tube first, and then carefully insert syringe B. Push in the plunger on syringe B. (The action of depressing the plunger on syringe B exerts pressure on the water, in the same way that water at the bottom of a lake is squashed down by the water above it.) What happens?

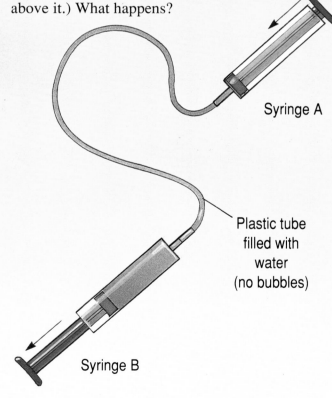

Syringe A

Plastic tube
filled with
water
(no bubbles)

Syringe B

THE TECHNOLOGY OF DAMS: FOUNDATIONS

Any big structure like a dam must stand on strong, solid foundations. If the valley is not made of solid rock, foundations have to be built. First the river must be **diverted** from its usual course, so that the building site can be kept dry. A low, temporary dam called a *cofferdam* stops the river water, which flows either through a channel by-passing the site, or through a specially built tunnel. Once the site is dry, work can begin.

If the ground is soft or loose, the builders have to dig down until they reach solid rock. Then strong foundations can be built. If the dam is an embankment, they may build a cut-off – a central **core** of clay to prevent water seeping through the dam. If much water is expected to seep through, they may add a layer of clay or concrete, called a run-off, under the dam, so that the water can run away instead of collecting and weakening the dam.

Concrete is a mixture of gravel, sand, cement and water, which can be moulded into almost any shape. When it has set hard, it is very strong against pushing forces, known as compression. **Reinforced concrete** has steel wires inside it to make it stronger. Sometimes the wires are stretched while the concrete is still wet.

When it dries, they try to pull back to their original length. This force **stresses** the concrete, making an extremely strong material.

SLUICES AND SPILLWAYS

The level of a reservoir is controlled by using sluices – pipes or channels fitted with gates which can be opened and closed. They also contain **screens** to stop plants, wood or rubbish getting through them. Some dams have sluices high up, letting water spray out in huge jets to the valley below. This prevents the water wearing away the base of the dam.

Spillways prevent water lapping over the top of the dam, which might eventually destroy it. If the reservoir rises too high up a sloping dam, the water escapes down a spillway, a chute ending in a '**ski-jump**', which throws the water out and stops it pounding the dam base. In straight or arch dams, a spillway pipe through the dam takes away any overflow.

The sluices on the Kariba Dam are positioned high up. When the sluices are opened the water jets out under high pressure.

3 Great dams

A ROCK-FILL EMBANKMENT: TARBELA

The Tarbela Dam in Pakistan, finished in 1975, is one of the biggest in the world. It is not the highest or longest, but it probably contains more material than the Great Wall of China, which was 3,200 km long!

The Dam was built to control the water in the River Indus, which changes with the seasons.

The Tarbela Dam was built by the simplest method there is. A special channel was made to divert the river. Then earth and rock were slowly piled 148 m high to make the dam.

The Indus at Skardu, upstream from the Tarbela Dam. In monsoon season these waters became a torrent, flooding down to the Dam.

HOW TO STOP DAMS LEAKING

What you need: A deep tray; sand or earth; a plastic bag or sheet.

What you do: Use the sand or earth to build a model embankment dam across the tray. Fill one side of the tray with water, and see how long it takes for water to start seeping through the dam.

Now rebuild the dam, with the plastic sheet lying under the reservoir and up through the middle of the dam. Try the experiment again. Now how long does it take for water to seep through the dam? Dam-builders use this idea when they build big embankment dams. But they use clay or concrete instead of plastic.

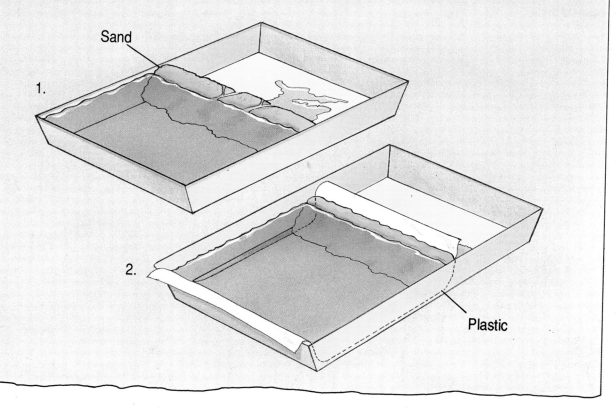

Sand

1.

2.

Plastic

The dam has a core of special material to stop it leaking. The main dam is 2.7 km long, and a reservoir stretches 80 km back behind it. Its water is used for irrigation, and to generate hydroelectric power.

A CONCRETE GRAVITY DAM: GRAND COULEE DAM

The Grand Coulee dams the River Columbia, USA. It stays in place because of its enormous weight — it is the world's biggest concrete dam.

It is a huge wall, 168 m high and 171 m long, with one steeply sloping side facing downstream. It has a road along its top, and a big hydroelectric station at one end. The Grand Coulee Dam was built by cutting off one part of the river at a time with a temporary dam. Builders dug down behind it until they reached solid rock. Nearly 20 million tonnes of concrete were added to the foundations to make the dam. Its reservoir is 240 km long.

The Grand Coulee Dam is not only the largest dam in the USA; it is also the world's largest concrete dam.

AN ARCH DAM: KARIBA

The Kariba stands in a narrow, rocky gorge on the mighty Zambezi River in Africa, between Zambia and Zimbabwe. It is a rounded wall of concrete 128 m high, held firmly in place by the water pressing it against abutments on the valley sides. Six sluices let water jet out of the middle of the wall, so that the water falls well clear of the dam's base. A road goes over the top. It was finished in

The Kariba Dam hydroelectric scheme generates most of Zambia's, and Zimbabwe's electricity.

1959, and now provides most of the electricity used in the two countries. Lake Kariba, the reservoir, is a huge 280 km long.

A BUTTRESS DAM: THE DANIEL JOHNSON

Buttress dams are like gravity dams, but the downstream side is held up by walls pushing into the valley floor below. The Daniel Johnson Dam in Quebec Province, Canada, was made from concrete blocks, and has fourteen buttresses linked by arches, with two huge buttresses in the middle. Finished in 1968, it is the largest dam of its type. It is 214 m high, and 1306 m long, storing enormous water reserves, and making a lot of electricity.

4 How dams help

WATER SUPPLY

Most of us do not worry about where our water comes from. We just turn on the tap and out it flows. But first, the water has to fall from the sky as rain or snow. It runs into lakes and rivers, and finally into the sea. Some of it **evaporates** into the air, where it forms clouds ready to rain again. This process is called the water cycle. But in some countries it rains only at certain times of year, or hardly ever. Dams help to store water for dry times, and water from reservoirs can be piped to fields and villages which would otherwise be far from water.

The sluices and spillways of dams are used to control the amount of water allowed through. Crops need much more water at some times of year than at others. But those times may not be when there is a lot of rain.

A constant water supply is perhaps our most valuable resource. Water sustains all life and human activity. Every day huge amounts are used for domestic use, in farming and in industry.

An aerial photo showing the full extent of the Uribante Capara Dam project in the Andes, Venezuela. The dam was specially built to generate much-needed hydroelectric power. Hydroelectric power is a clean form of energy and does not pollute the environment.

Left These turbines, used to generate hydro-electric power, are situated within the Hoover Dam's wall.

Below When there is danger of flooding, the Thames Barrier on the River Thames in London, UK, swings ten gates into position which can stop the water.

Irrigation schemes using dams, reservoirs, canals and channels, can make sure that enough water is in the right place at the right time.

Dams can also let water through in dry weather to stop the river below drying up, or can hold it back to prevent flooding of the valley below.

If a reservoir is needed to supply people living and working in a town or city, and its schools, factories and power stations, the reservoir should be full all the time. It may not need a dam – sometimes a simple walled reservoir or a natural lake will do.

WATER POWER

Hydroelectric power is generated from the energy in falling water. Water from behind the dam rushes down pipes into the huge **generator**, crashing with great force on to the blades of a **turbine.** This makes the turbine spin, driving a generator round to produce electricity.

In the Snowy Mountains of New South Wales, Australia, the waters of three rivers are diverted by aqueducts and tunnels, and harnessed by sixteen dams and seven hydroelectric power stations to make electricity, and for irrigation. The huge project is controlled from one central control room. Hydroelectric power is a clean form of power, and does not **pollute** the **environment.** Water is also a readily

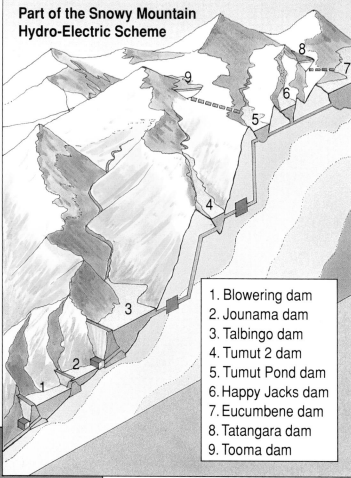

Part of the Snowy Mountain Hydro-Electric Scheme

1. Blowering dam
2. Jounama dam
3. Talbingo dam
4. Tumut 2 dam
5. Tumut Pond dam
6. Happy Jacks dam
7. Eucumbene dam
8. Tatangara dam
9. Tooma dam

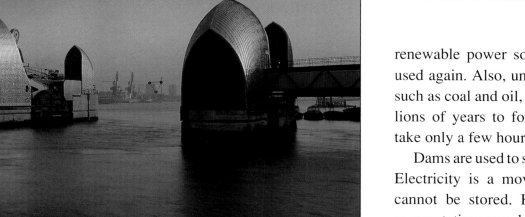

renewable power source. It can be used again. Also, unlike fossil fuels such as coal and oil, which take millions of years to form, rain clouds take only a few hours to form.

Dams are used to store power, too. Electricity is a moving force, and cannot be stored. However, some power stations use electricity to pump water uphill, back into a reservoir.

This is often done at night, when electricity demand is low. The water falls again to re-generate electricity.

CLEANER WATER

River water, washed down from the hills, contains tiny specks of silt and dirt. In the calm, still waters of a reservoir, this silt slowly sinks to the bottom, leaving the water cleaner. However, the water still needs to be treated and **filtered** before it is clean enough to drink.

WATER BARRIERS

Some dams are not built across rivers, but beside them or along coastlines. These dams are built to stop very high tides or rivers flooding the land. High mud banks called levees are built along rivers that lie in very shallow valleys. The Mississippi in the USA is lined by 6,000 km of levees.

The Netherlands is flat, and much of it lies below sea-level. The sea is kept out by wall dams called dykes. The Dutch have also **reclaimed** land from the sea. They built dykes across a bay, and then pumped out the seawater.

Many cities are built near the sea or by large tidal rivers. Storms and unusually high tides can cause flooding. London, a port on the tidal River Thames, has been flooded in the past when seawater has surged up the river.

Dams may stop fish swimming up and down a river. Salmon, for instance, swim upstream to breed, before returning to the sea. So engineers have thought of a way to help them, by building fish ladders. A series of pools, linked by sloping pipes, let fish swim around the dam.

Now it is protected by a huge flood barrier. It is 520 m long, and has ten gates which normally lie flat on the riverbed so that ships can pass. If there is a danger of flooding, the gates swing upwards to stop the water.

5 Dam drawbacks

The Temple of Philae, rescued from flooding when the High Aswan Dam was built.

Dams are important to us all. But they also have their drawbacks. Big dams are the most massive structures on Earth. They can be very expensive, too. In some dry countries, where water supply is a great problem, dams would help. But these countries are too poor to build the dams they need without help from richer nations. People living in these areas have to fetch and carry all their water from wells, or from rivers.

Dams change the landscape for ever. Damming a small stream may flood a farmer's field. Damming a valley can cause great disruption. Whole farms and villages – houses, schools, public buildings – are drowned as the reservoir slowly fills, and end up deep underwater. When the High Aswan Dam was built over the Nile in Egypt in the 1960s, many ancient monuments and temple statues had to be rescued from the rising waters and moved to safety.

When the Kariba Dam was built, and the Zambezi River rose to form Lake Kariba, over 30,000 tribespeople had to be moved to new homes. Animals can suffer too. As the Zambezi valley flooded, thousands of wild animals became trapped on little islands, and had to be rescued by boat in Operation Noah. On the other hand, the new lake now provides fish to feed local people.

SILT

As you have seen, silt in reservoirs settles at the bottom, leaving the water cleaner. But that is not always a good thing. The silt can build up and clog a reservoir, until it cannot hold enough water. Also, the silt and soil in water is good for the land around a river. It makes the valley **fertile**. The Nile used to flood the Aswan valley every year, enriching the fields with silt. Now the river is dammed, this annual enrichment no longer takes place, so the soil is less fertile. This means the land produces less food than it did before, making life harder for farmers.

EARTHQUAKE, LANDSLIDE AND FLOOD

If the engineers have done their sums and preparation well, a dam will be strong enough to hold back the water behind it, whatever happens. But tragic accidents still happen, especially because of the unimaginable power of natural forces.

Areas of rain-forest were drowned as the Tucurui Reservoir in Brazil slowly filled up.

Very rarely, a dam weakens before it bursts. In 1976, the Teton Dam in Idaho, USA, was new and its reservoir was filling up, when water began trickling through a weak spot in the 95 m high dam wall. Soon the trickle grew to a flood as the dam burst. There was enough warning to move most people away from below, and the dam burst was filmed for television! Despite the warning signs, several people died, and the water and sand pouring down the valley caused immense damage.

An even worse disaster happened

A film still from **Superman, the movie,** *showing a mock-up of a dam bursting. This picture gives some idea of the spectacular damage caused when a dam bursts.*

in western India in 1979, when an earth dam collapsed under the pressure of floods. 5,000 people were killed, and many more made homeless.

Even when a dam is strong, the rock around it may not be as strong as was thought. In Fréjus, France, in 1959, the rock holding the abutments of an arch dam gave way, loosening the dam. The dam collapsed, killing over 300 people.

Earthquakes can also loosen a dam's foundations, or even crack the dam itself. Sometimes the weight of the

HOW A DAM BREAKS

What you need: A deep plastic tray; sand or earth; a pencil; water.
What you do: Preferably outside, make a dam across the tray as before, and fill one side with water. When the 'reservoir' is full, make a small hole near the top of the dam with your finger or a pencil. What happens? Can you see how a small crack can make a dam burst, and why so much damage is done when it does burst?

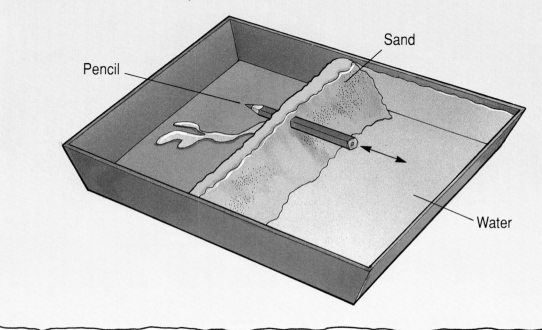

Pencil

Sand

Water

water in a big reservoir can actually cause an earthquake or a landslide, because the rock is not used to the weight. In 1963, the weight of water behind one of the world's highest dams, at Vaiont in Italy, caused a mountainside to fall into the water. A huge wave of water swept 100 m over the top of the 262 m-high dam, crashing into the valley below. It killed over 2,000 people, but the dam – a strong arch – survived almost unharmed.

The worst dam disaster happened over a century ago in the USA. A huge dam in Philadelphia burst in 1889, tossing houses together like eggshells. 10,000 people died.

Whenever a disaster happens, engineers look carefully to see why it happened. They learn their lessons from the tragedy, and take great care to make sure it doesn't happen again.

6 Thinking about dams – the story of the Yangtze

Like planners of any large structure, the planners of a dam have to decide whether the benefits will be greater than the problems it may cause. How much will it cost? Who will pay for it? Will it create more fertile land and allow crops to grow where none could grow before? Will it produce electricity? Will it control flooding? How many people will have to be moved to new homes?

All these questions must be answered before the designers and engineers can start work on the dam itself. And sometimes the answers are not easy.

THE WORLD'S BIGGEST DAM

China's Yangtze River is the longest river in Asia. It is 5,525 km long. On its way down from the high mountains of Tibet to the sea, it flows across the great plains of central and eastern China, which produce nearly half of all the crops grown in China. The river basin also contains several large cities, including Nanking, Wu-han and Shanghai.

The problem is that the river and its **tributaries** often spill over their banks and cause large floods. This is because of the large amount of silt carried by the river. When the river slows down (as it leaves the mountains and starts crossing the plains), some of the silt it carries sinks to the riverbed. As a result, the river, which is held in by dykes, is higher in some places than the surrounding plain. Every few years disastrous floods occur, such as the great floods of 1931 and 1954. The 1931 flood left 40 million people homeless. In July 1991, near Shanghai, an area of farmland nearly as big as England and Scotland was flooded. Over 2,000 people died, and a million were left homeless.

WHAT TO DO?

The idea of damming the Yangtze to control flooding was first suggested in the 1920s. After 30,000 died in the 1954 floods, detailed plans were made. However, the idea was forgotten, until the recent floods made the Chinese think of it once more.

The suggested place for the dam is in the Shangsia Gorges in central China. Here the river passes through three beautiful gorges – deep, narrow valleys with steep, rocky sides – where the river is as deep as 200 m, making it the world's deepest river.

The proposed dam would be the biggest dam project in the world. It is not just designed to prevent flooding of the plains below. It would also generate twice as much hydroelectric power as the largest existing hydro-electric dam, in South America.

The farming plains on the banks of the Yangtze River, China, are enriched with the silt when the river floods. Damming the river will stop this.

Its 26 generators would produce one-sixth of China's electricity.

On the other hand, the dam would flood 600 km of the Yangtze valley, making a lake nearly twice the size of the Isle of Wight. It would cost a huge amount of money. Over a million people would have to be moved. Some people think that the dam would be an environmental disaster which would upset the balance of nature in the three gorges – one of China's most famous areas of natural beauty.

What is worse, the engineers do

not agree that it will work. Some leading engineers doubt that it would prevent flooding, because flood waters often reach the plains from large tributaries that join the Yangtze below the gorges. In fact, they say it could make flooding worse. If the river's silt was trapped behind the dam, the river could wear away the dykes and banks on the plains. The lack of silt would also make the plains less fertile. Any break in the dam, caused by an earthquake or landslide, or even by war, could result in a flood wave that could kill millions.

In March 1992, after nearly seventy years of hesitation, the Chinese parliament voted to build the dam. If you had been a member of that parliament, how would you have voted?

Some engineers argue that damming the Yangtze will put a stop to the disastrous floods which occur every few years, causing serious damage.

Glossary

Aqueducts Bridges built to carry flowing water from one place to another.

Artificial Made or caused by people, not by nature.

Brace Give extra strength, firmness and support by using the solid ground to hold up the dam wall.

Buttress A support used in large buildings like cathedrals as well as dams. The buttress props up a wall from the side, holding it up.

Compressing Squashing and pressing with great force.

Concrete A mixture of sand and gravel held together with a glue of cement and water. It sets very hard as it dries.

Core A waterproof wall or layer in the middle of the embankment.

Diverted Made to go another way.

Drought A time of dry weather, when there is not enough water for people, their animals and their crops or gardens.

Engineers People who build complicated structures or machines.

Environment Everything around us (and all living things) – the air we breathe, the land we live on, the water we drink, and so on.

Evaporates Changes from liquid into a gas or vapour, which becomes part of the air.

Fertile Full of rich soil, able to produce a lot of good crops.

Filtered Passed through a fine mesh, like a sieve, to remove any remaining particles.

Foundations The solid base on which a dam (or any structure) stands.

Geological surveys Scientific studies of the rocks and ground in the place where the dam is to be built. They often involve drilling deep into the ground to take samples of rock and soil.

Generator A machine for making electricity.

Gravity The natural force which attracts all things to each other. The Earth is very big, so it pulls us firmly down to the ground – otherwise we would float away.

Greenhouse effect In a glass greenhouse, sunlight comes through the glass and is taken in by the plants. They release the energy they do not use as heat. This heat energy does not go through the glass. It is trapped, making the greenhouse warmer. Exactly the same thing is happening in the air around the Earth. Some gases in the air let light from the sun through, but not heat coming back from the planet. We are making more of these gases which trap heat. This is making the Earth get warmer.

Harnessed Taken and used for the benefit of people.

Irrigation Making crops grow better by keeping them watered.

Leisure activities All the things people do in their spare time, such as sports.

Pollute Release substances such as waste gases, liquids and solids which poison our surroundings (the environment), and harm or even kill living plants or animals.

Reclaimed Got back what was useless, and made it useable.

Reinforced Made even stronger.

Resource A natural supply of a useful material or substance.

Screens Meshes or grilles of metal which let water through, but catch most solid objects. They have to be taken out regularly and cleaned.

Ski-jump Named after the slope used by ski-jumpers, a ski-jump is a long steep slope which turns upwards just at the bottom. It throws the water (or ski-jumper) upwards and out a long way.

Sluices Equipment for controlling the flow of water through dams. Usually they are solid gates which slide open and shut in special frames, placed in a channel somewhere below the top of the dam.

Stresses Pulls and squashes the material.

Tributaries Side rivers which flow into a main river.

Turbine A machine in which the energy of water or steam is used to turn a shaft or axle. It looks something like a ship's propeller.

Upstream Uphill, against the flow of the water which runs downstream.

Further reading

Structures by Malcolm Dixon (Wayland, 1990)

Building Technology by Mark Lambert (Wayland, 1991)

Dams by Neil Ardley (Macmillan, 1989)

Energy Without End: The Case for Renewable Energy by Michael Flood (Friends of the Earth, 1986)

Renewable Energy, The Power to Choose by Daniel Doudney and Christopher Flavin (Nathan, 1983)

Index

Numbers printed in **bold** refer to pictures as well as text.